Why are Plants Wonderful?

A Coloring Book

By Michael Reed

Why are Plants Wonderful?

A Coloring Book

By Michael Reed

Copyright©2017 by MR

All rights reserved.

Published in Chicago, IL by MR

Acknowledgement

I thank God for giving me the interest and knowledge for this book.

Plants come in many forms, from the moss and liverwort to the biggest trees on Earth.

Liverwort

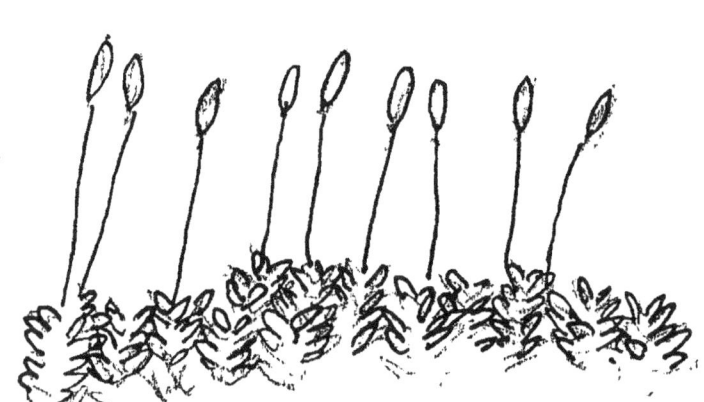

Moss

Fern spp.

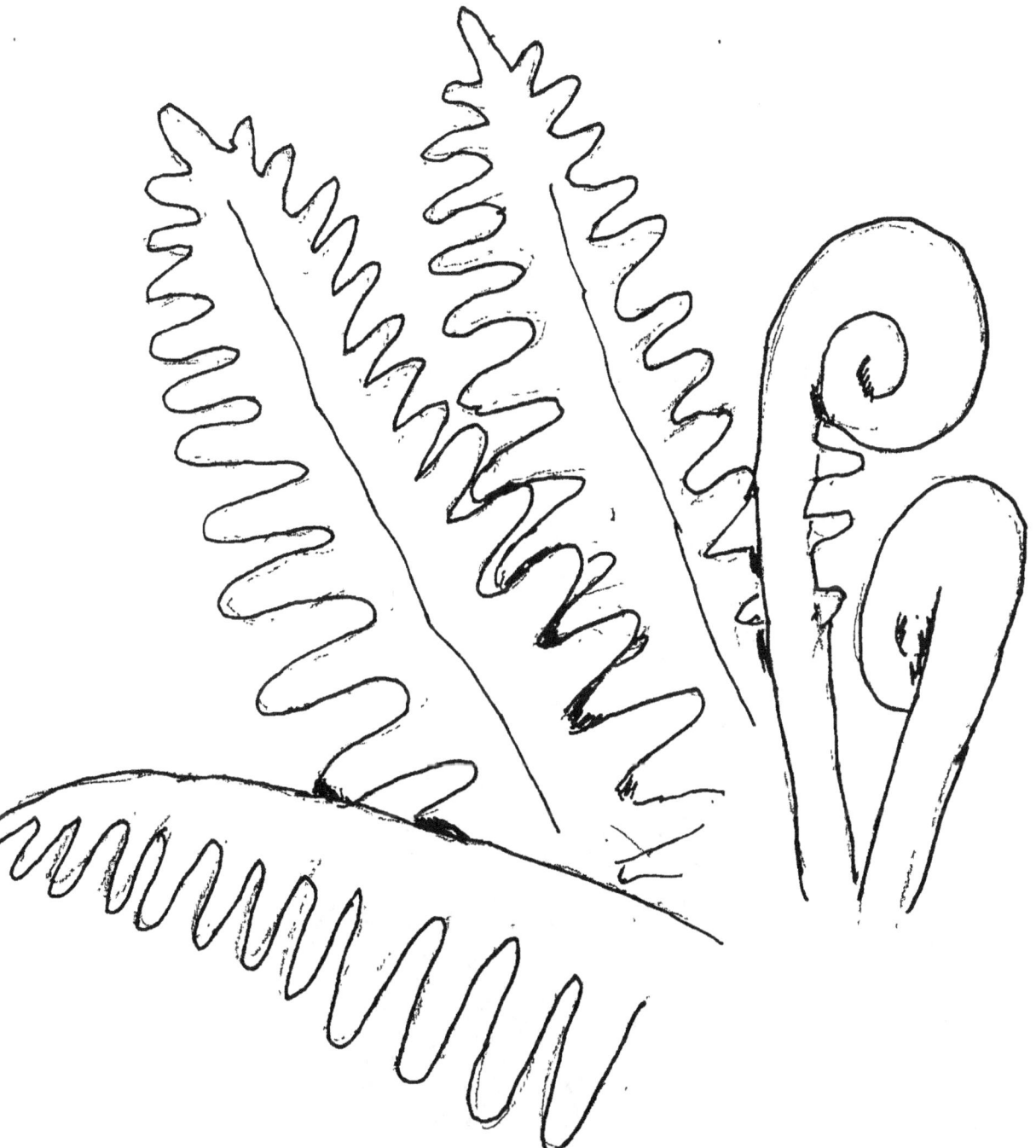

Tree fern

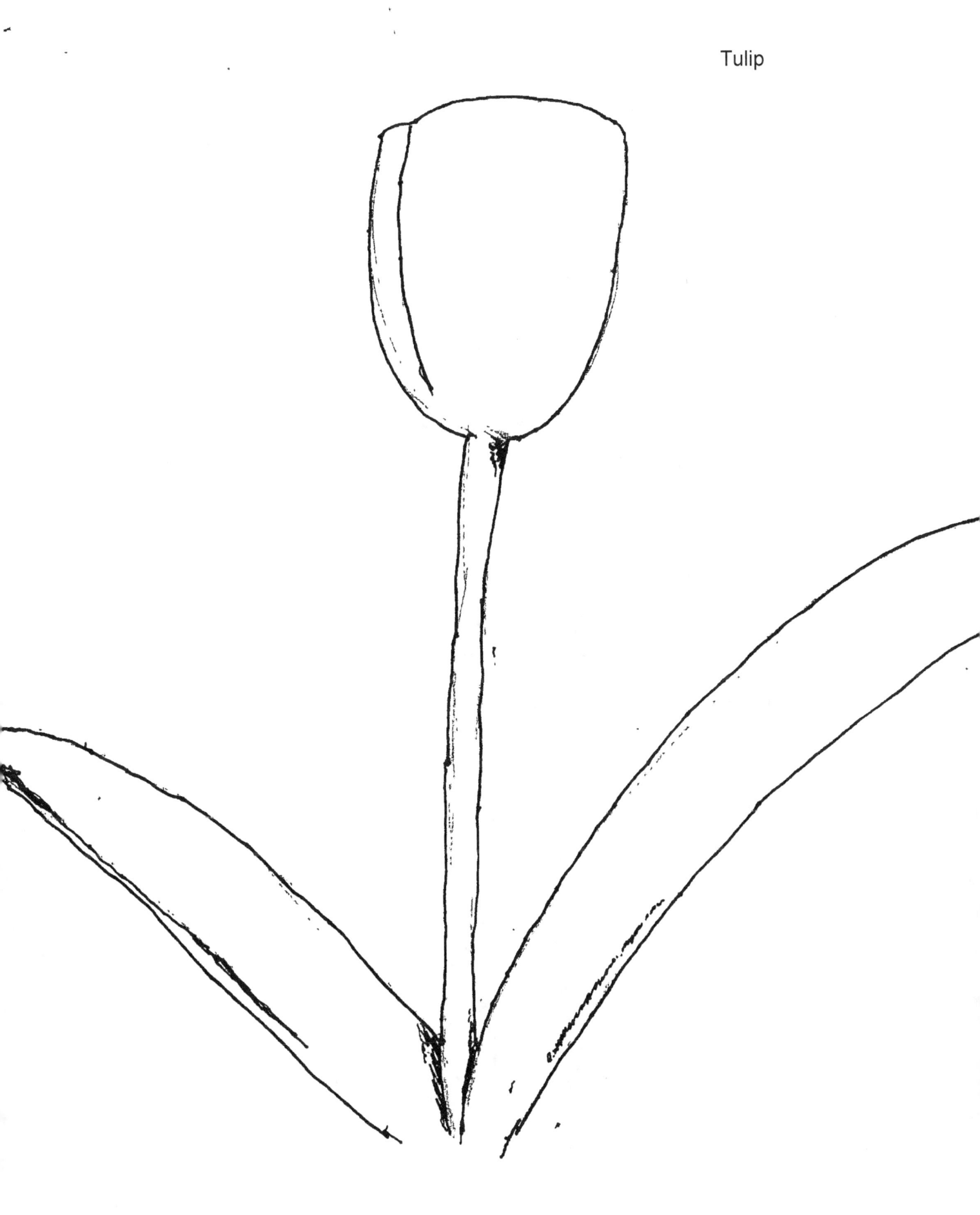
Tulip

Coconut Palm
(*Cocos nucifera*)

Horsetails

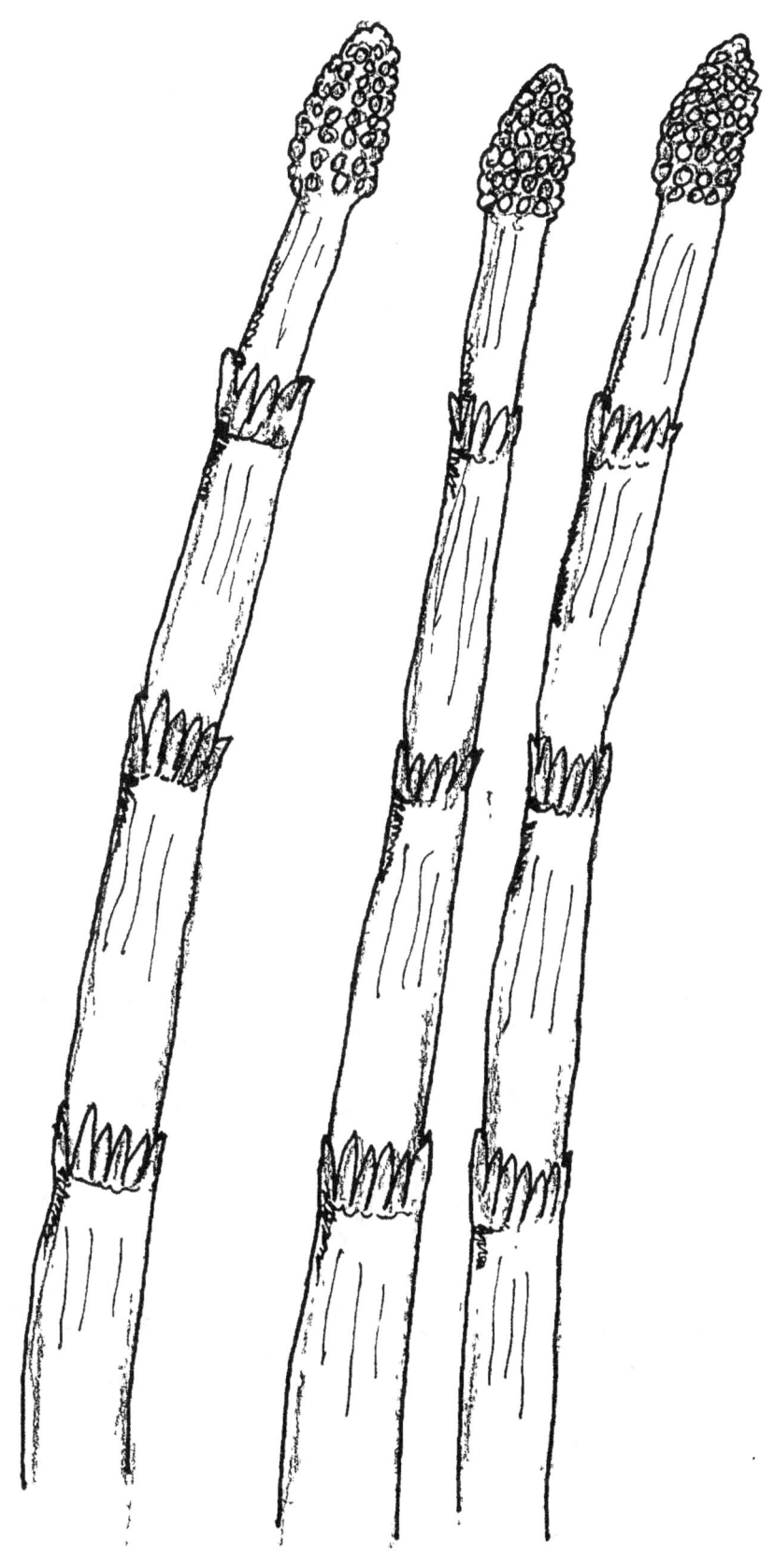

Bromeliads

Giant sequoia forest

(*Sequoiadendron giganteum*)

Plants are some of the most wonderful organisms on Earth because...

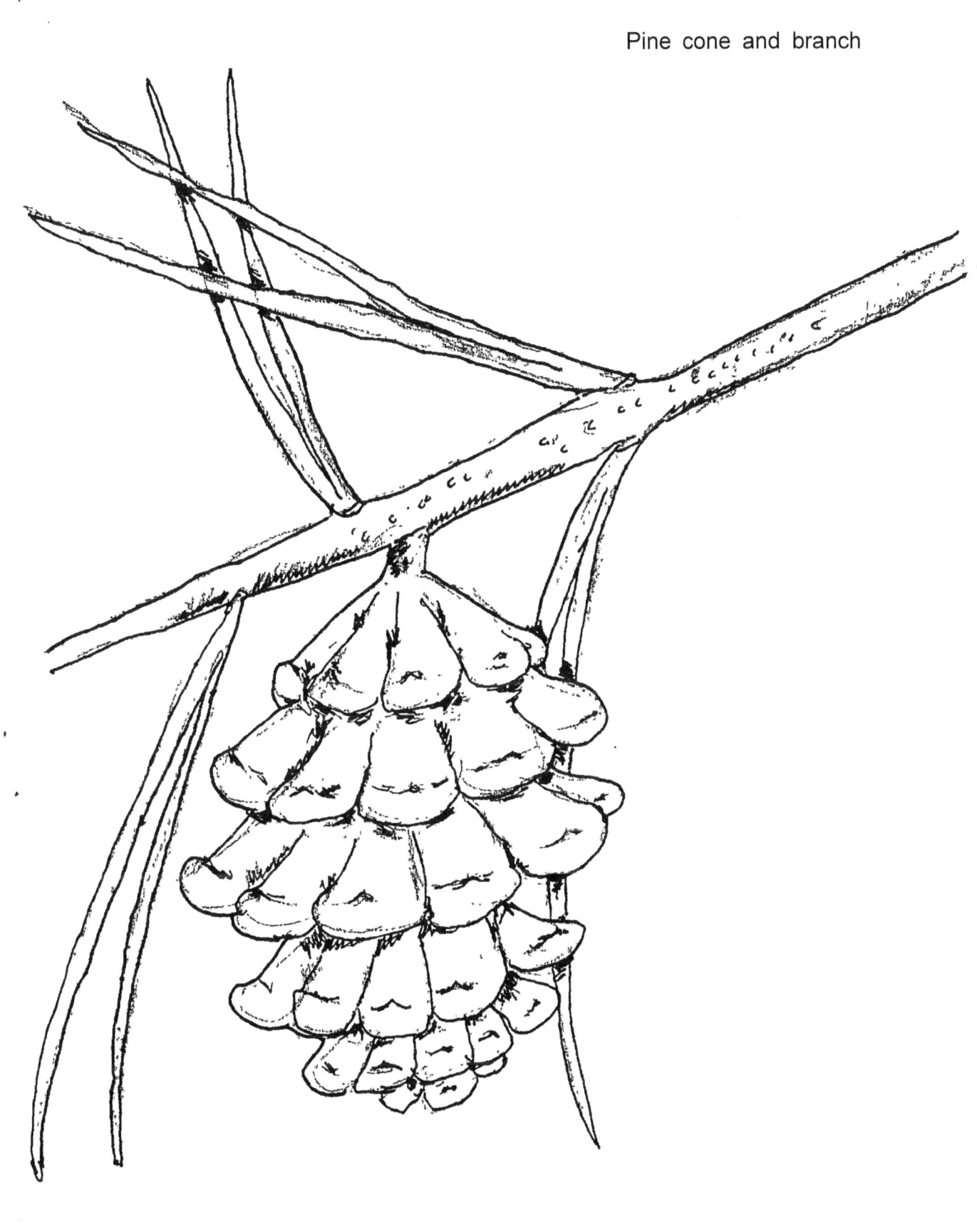

Pine cone and branch

...they help to provide the support for many habitats, from the tundra to the tropical rainforest, on Earth.

Tundra

Boreal forest

Bald Cypress swamp in the Southeastern USA
(*Taxodium distichum*)

Cacti in the American Southwest

Canopy of a tropical rainforest

Plants are useful because they...

Sunflower

... use sunlight to help to turn carbon dioxide and water into oxygen and energy.

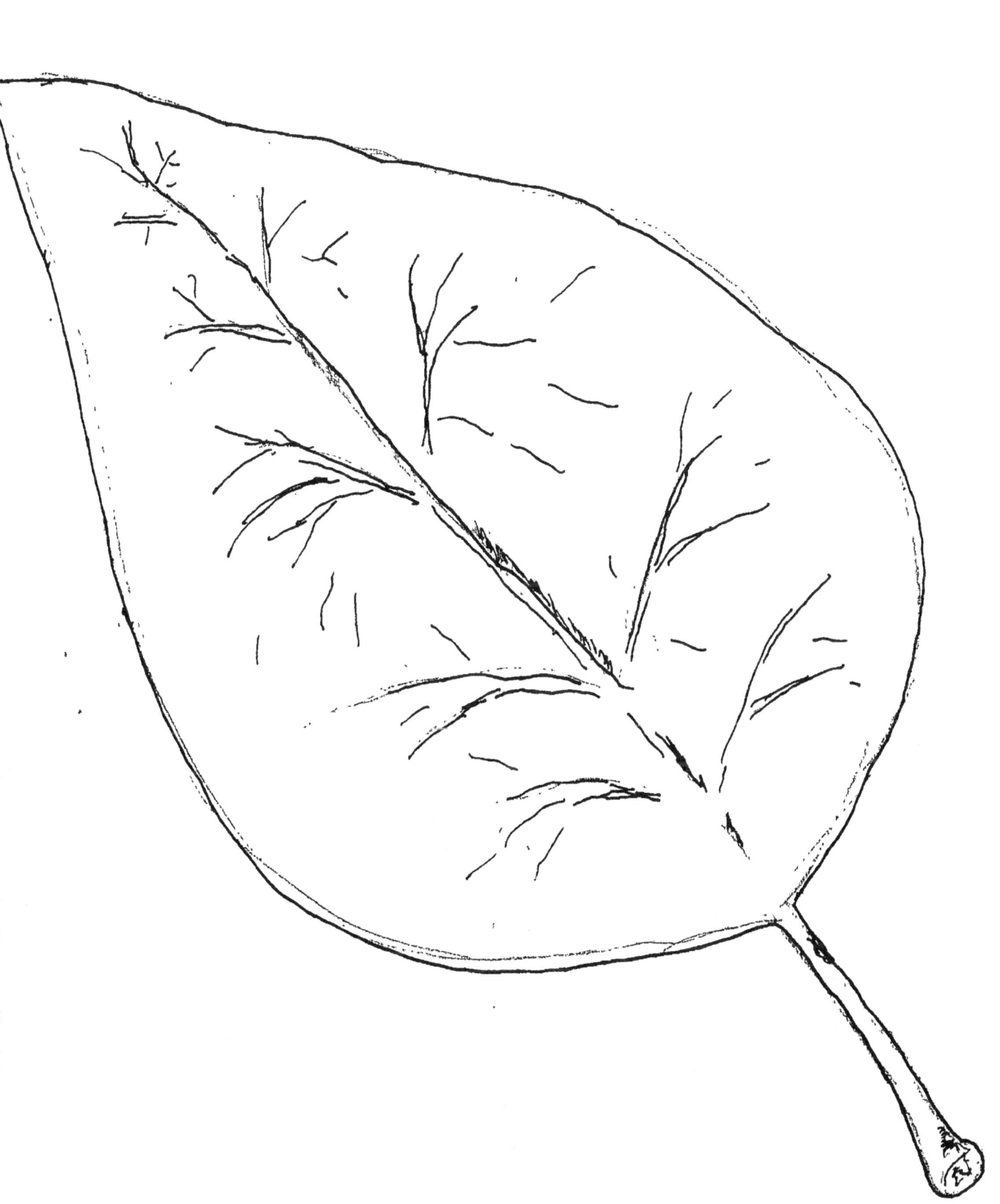

Leaf of a Pitcher Plant

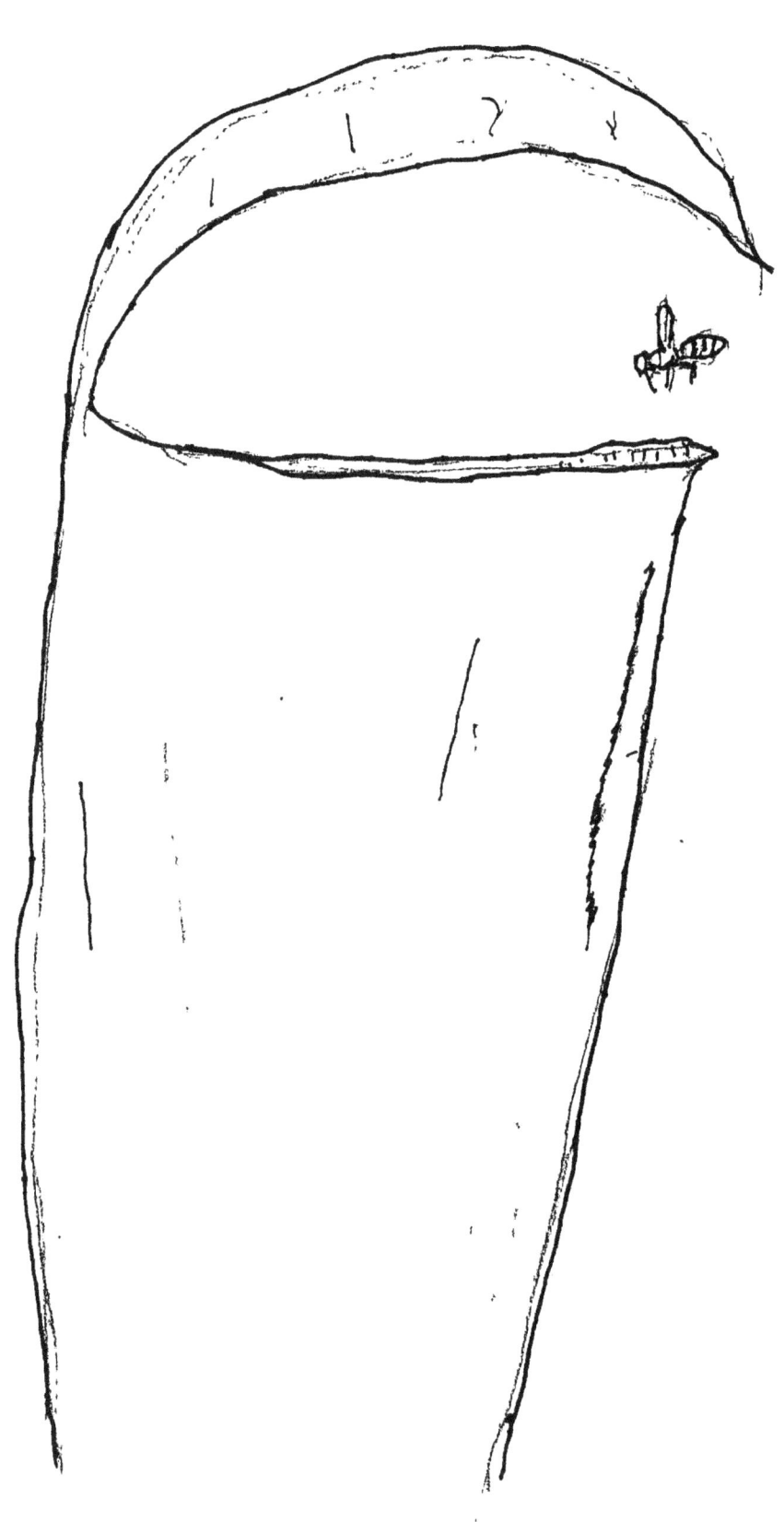

... help to produce much of our food

Apples

POPCORN

... help to provide compounds for our medicines

Madagascar Periwinkle

(*Vinca rosea*)

Compounds: vincristine & vinblastine for anticancer treatment

... and materials for our clothes

Cotton

(*Gossypium species*)

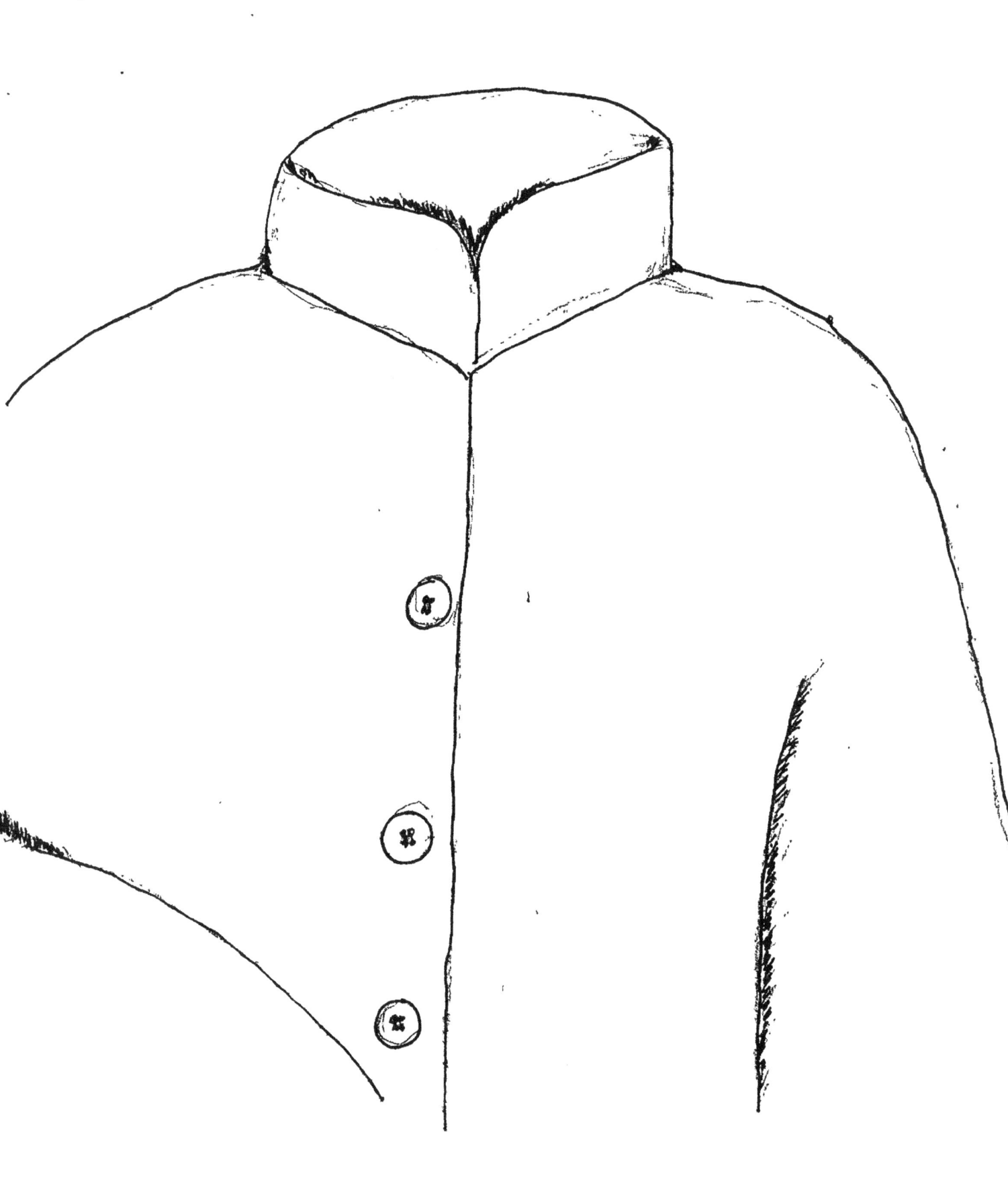

We can learn more about plants from our teachers, museums, books, and the computer. However, we can turn to God to get understanding about them.

FOOD PLANTS OF THE WORLD

BOTANY

References for Use

Brockman, Frank C. Rebecca Merrilees, Herbert S. Zim, Ed. **Trees of North America.** Golden Press, New York. 1986.

Ennos, Roland & Elizabeth Sheffield. **Plant Life**. Blackwell Science, Inc. Malden. 2000

Marinell, Janet Ed.in Chief. **Plant: The Ultimate Visual Reference to Plants and Flowers of the World.** DK Publishing, Inc New York. 2005.

Mauseth, James E. **Botany: An Introduction to Plant Biology**. Jones & Bartlett Learning, Burlington. 2017.

Sibley, David Allen. **The Sibley Guide to Trees.** Alfred A. Knopf, New York. 2009.

www.ingramcontent.com/pod-product-compliance
Lightning Source LLC
Chambersburg PA
CBHW062207220526
45470CB00009B/2958